DISCARD

Ocean Animals
Walruses

by Christina Leaf

BELLWETHER MEDIA • MINNEAPOLIS, MN

BLASTOFF! Beginners

Blastoff! Beginners are developed by literacy experts and educators to meet the needs of early readers. These engaging informational texts support young children as they begin reading about their world. Through simple language and high frequency words paired with crisp, colorful photos, Blastoff! Beginners launch young readers into the universe of independent reading.

Sight Words in This Book

and	find	it	the	two
are	for	like	their	use
at	have	long	them	water
big	help	look	these	
called	in	on	they	

This edition first published in 2022 by Bellwether Media, Inc.

No part of this publication may be reproduced in whole or in part without written permission of the publisher. For information regarding permission, write to Bellwether Media, Inc., Attention: Permissions Department, 6012 Blue Circle Drive, Minnetonka, MN 55343.

Library of Congress Cataloging-in-Publication Data

Names: Leaf, Christina, author.
Title: Walruses / by Christina Leaf.
Description: Minneapolis, MN : Bellwether Media, Inc., 2022. | Series: Blastoff! beginners: Ocean animals | Includes bibliographical references and index. | Audience: Ages 4-7 | Audience: Grades K-1
Identifiers: LCCN 2021001481 (print) | LCCN 2021001482 (ebook) | ISBN 9781644874837 (library binding) | ISBN 9781648343919 (ebook)
Subjects: LCSH: Walrus--Juvenile literature.
Classification: LCC QL737.P62 L43 2022 (print) | LCC QL737.P62 (ebook) | DDC 599.79/9--dc23
LC record available at https://lccn.loc.gov/2021001481
LC ebook record available at https://lccn.loc.gov/2021001482

Text copyright © 2022 by Bellwether Media, Inc. BLASTOFF! BEGINNERS and associated logos are trademarks and/or registered trademarks of Bellwether Media, Inc.

Editor: Amy McDonald   Designer: Laura Sowers

Printed in the United States of America, North Mankato, MN.

Table of Contents

Walruses!	4
Body Parts	8
Water and Land	16
Walrus Facts	22
Glossary	23
To Learn More	24
Index	24

Walruses!

Look at those big teeth!
Hello, walrus!

5

Walruses swim in the ocean. They rest on land and ice.

7

Body Parts

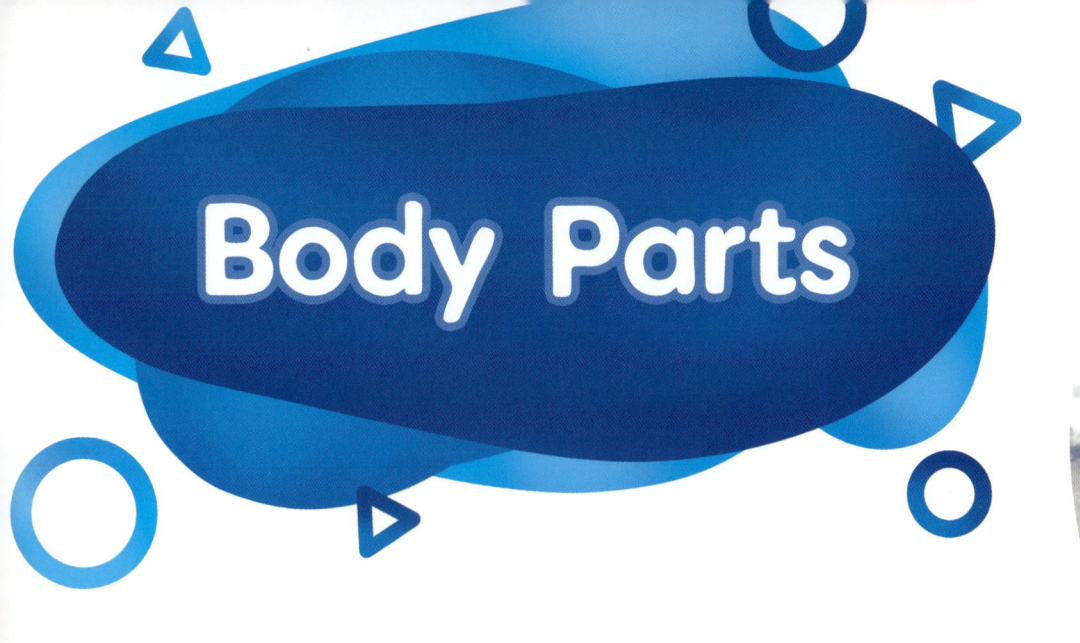

Walruses have long teeth. They are called **tusks**.

tusks

9

They have **whiskers**.
They feel for food.

whiskers

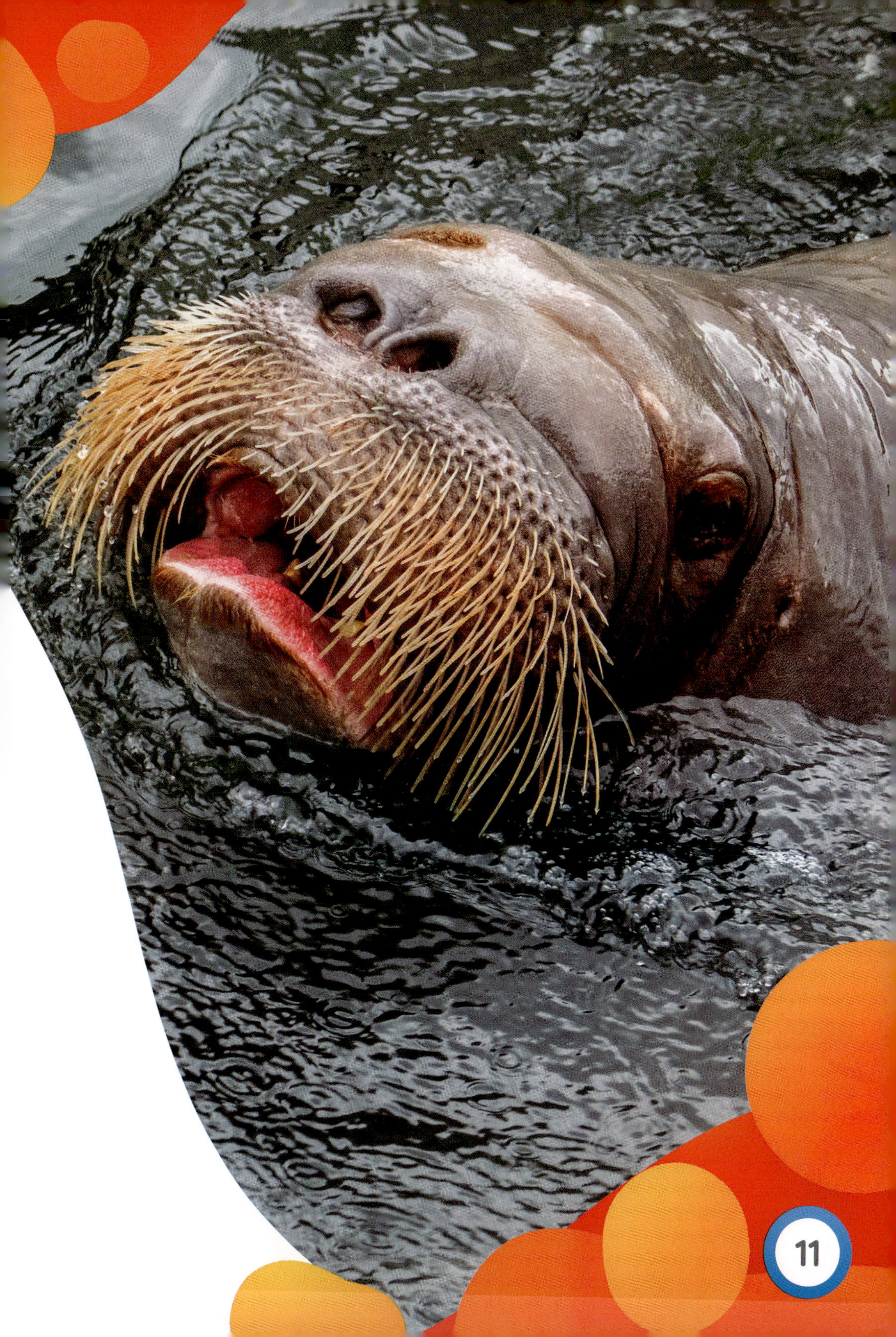

They have **flippers**. These help them swim and walk.

flippers

They have **blubber.**
It keeps them warm in icy water.

15

Water and Land

Walruses find food in water. They like clams and crabs.

crab

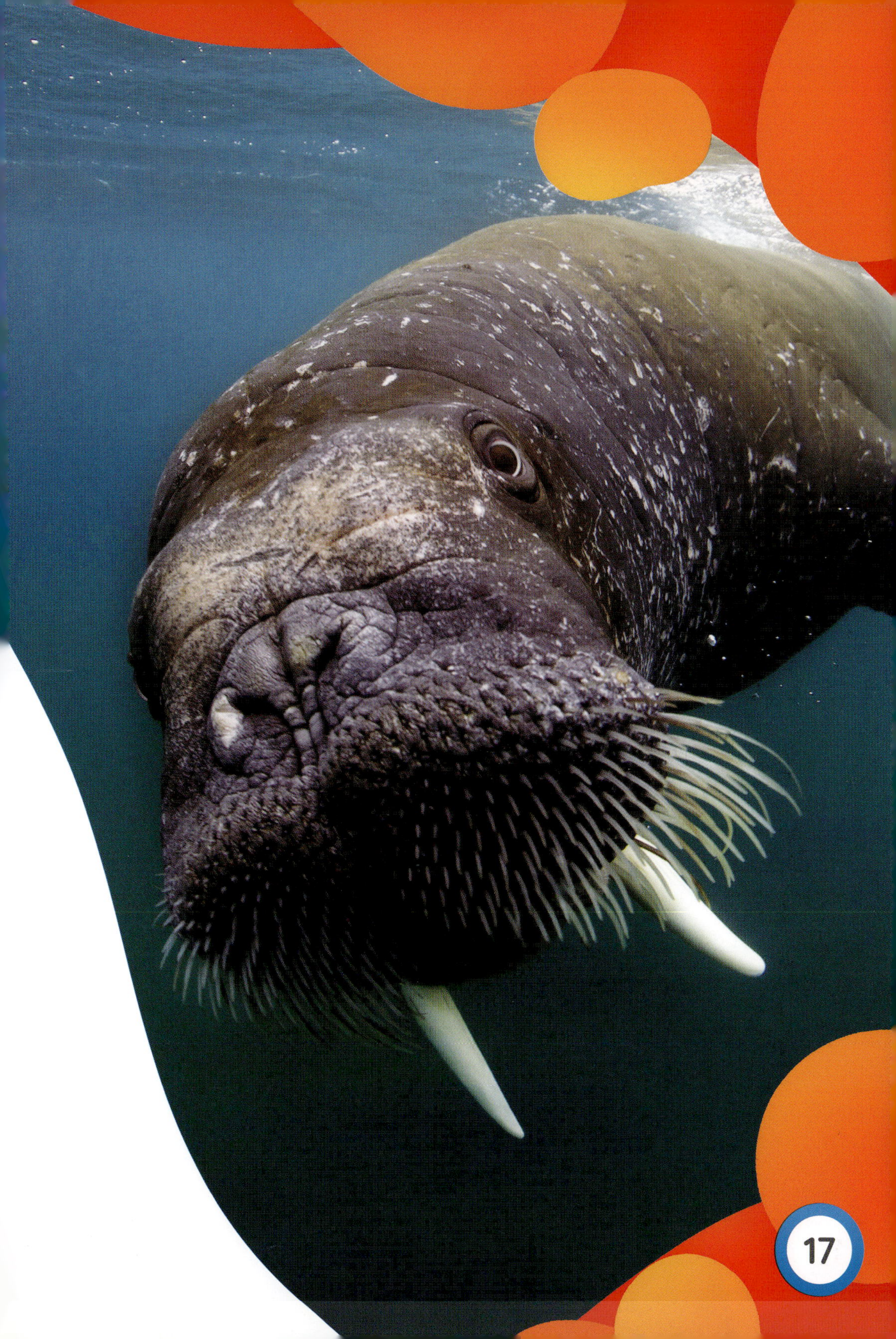

They live in big groups on land.

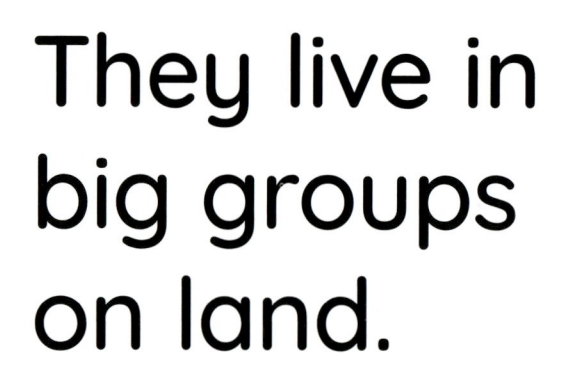

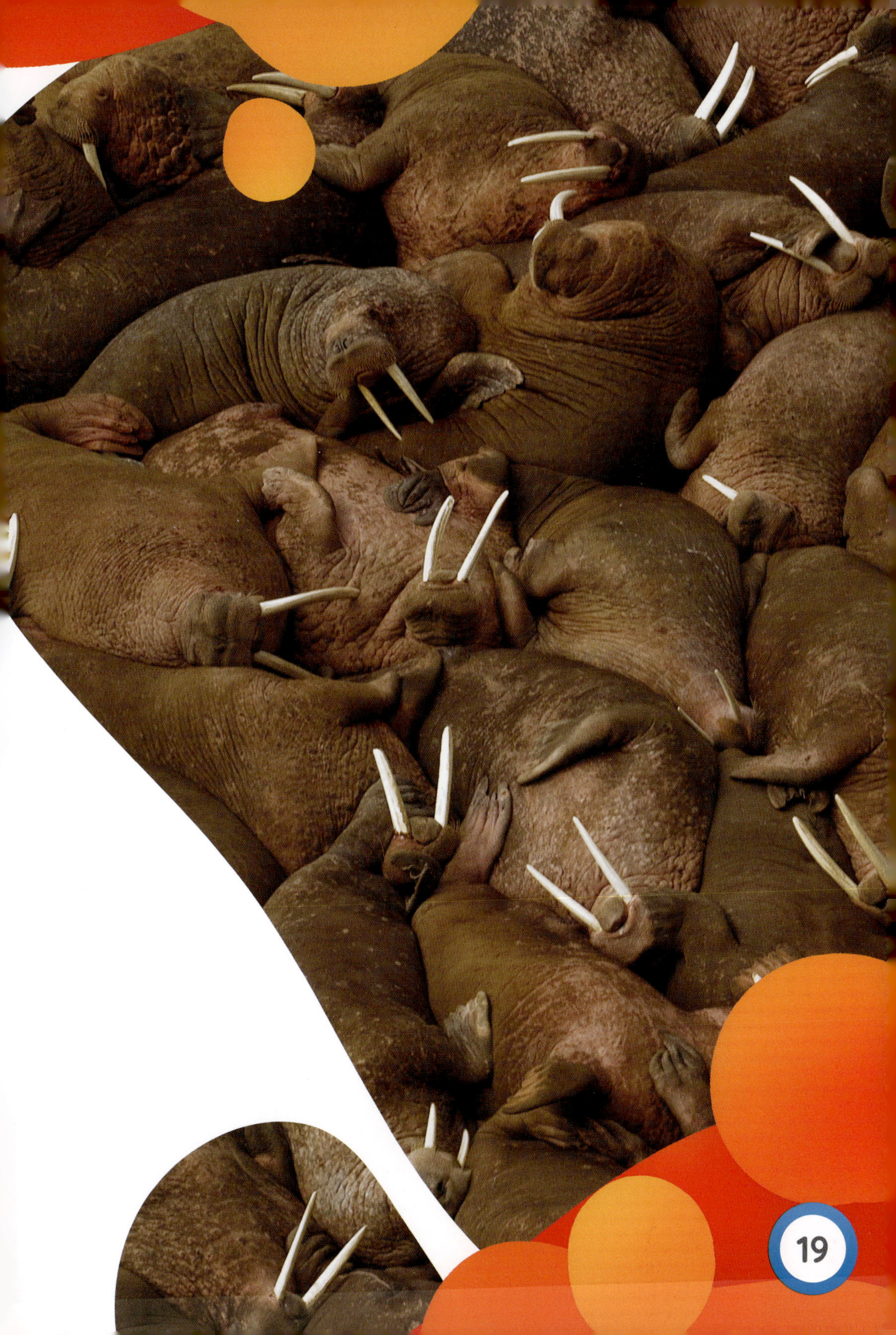

These two fight. They use their tusks. Ouch!

21

Walrus Facts

Walrus Body Parts

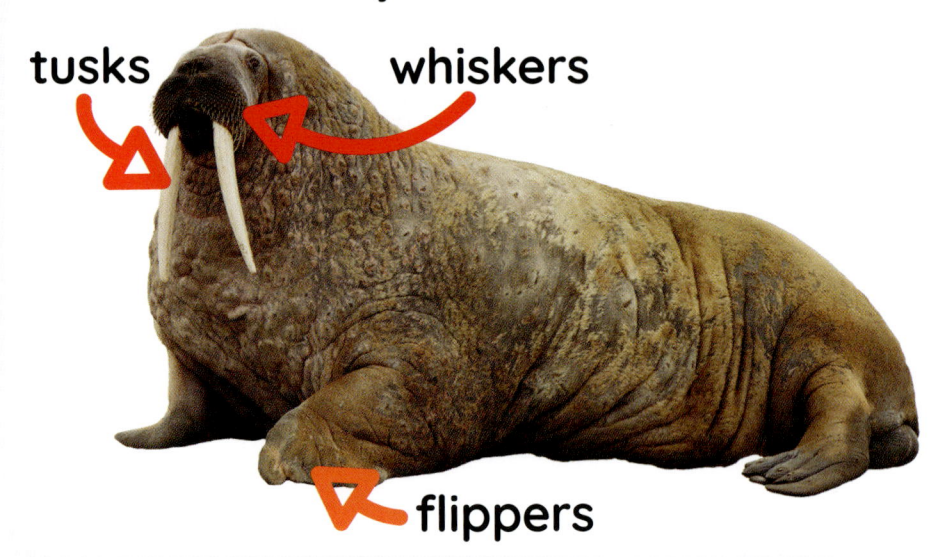

tusks　　whiskers

flippers

Walrus Food

clams　　crabs　　snails

Glossary

blubber

body fat that keeps walruses warm

flippers

flat body parts on a walrus

tusks

long teeth that stick out of a walrus's mouth

whiskers

hairs on a walrus's face

To Learn More

ON THE WEB

FACTSURFER

Factsurfer.com gives you a safe, fun way to find more information.

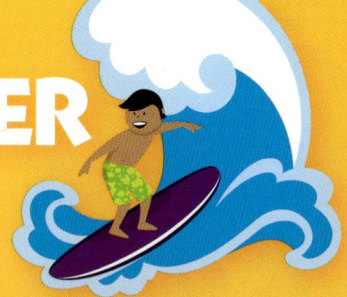

1. Go to www.factsurfer.com.

2. Enter "walruses" into the search box and click 🔍.

3. Select your book cover to see a list of related content.

Index

clams, 16
crabs, 16
blubber, 14
fight, 20
flippers, 12, 13
food, 10, 16
groups, 18
ice, 6, 14
land, 6, 18

ocean, 6
rest, 6
swim, 6, 12
teeth, 4, 8
tusks, 8, 20
walk, 12
warm, 14
water, 14, 16
whiskers, 10

The images in this book are reproduced through the courtesy of: Mikhail Cheremkin, front cover, pp. 12-13; Vladimir Melnik, p. 3; Zaruba Ondrej, pp. 4-5; Mats Brynolf, p. 6; jo Crebbin, pp. 6-7; Pascal Halder, p. 8; Egor Vlasov, pp. 8-9, 23 (flippers); Aleksei Verhovski, pp. 10, 22 (whiskers); Philippe Clement, pp. 10-11; ginger_polina_bublik, pp. 14-15; Nick Kashenko, p. 16; Biosphoto/ Alamy, pp. 16-17; Hal Brindley, pp. 18-19; Blue Planet Archive/ Alamy, pp. 20-21; Vladimir Melnik, p. 22; T-I, p. 22 (clams); Kondratuk Aleksi, p. 22 (crabs); valda butterworth, p. 22 (snails); Lillian Tveit, p. 23 (blubber); BMJ, p. 23 (tusks).